ACCESS TO MATH

AF886309

PROBABILITY

GLOBE FEARON EDUCATIONAL PUBLISHER
Upper Saddle River, New Jersey
www.globefearon.com

Executive Editor: Barbara Levadi
Editors: Bernice Golden, Lynn Kloss, Bob McIlwaine, Kirsten Richert, Tom Repensek
Production Manager: Penny Gibson
Production Editor: Walt Niedner
Interior Design: The Wheetley Company
Electronic Page Production: The Wheetley Company
Cover Design: Pat Smythe

Reviewers:
Beth C. Klein, B.S., Mathematics Teacher
Metuchen High School, Metuchen, NJ

Elizabeth Marquez, B.A., M.A.
Mathematics Teacher
North Brunswick Township High School,
North Brunswick, NJ

Copyright ©1996 by Globe Fearon Educational Publisher, a Division of Simon and Schuster, 1 Lake Street, Upper Saddle River, New Jersey 07458. All rights reserved. No part of this book may be reproduced or transmitted in any form or by any means electrical or mechanical, including photocopying, recording, or by any information storage and retrieval system, without permission in writing from the publisher.

Printed in the United States of America 3 4 5 6 7 8 9 10 99

ISBN 0-8359-1568-9

GLOBE FEARON EDUCATIONAL PUBLISHER
Upper Saddle River, New Jersey
www.globefearon.com

Textbook And Beyond
www.textbooknbeyond.com

CONTENTS

TO THE STUDENT

Access to Math is a series of 15 books designed to help you learn new skills and practice these skills in mathematics. You'll learn the steps necessary to solve a range of mathematical problems.

LESSONS HAVE THE FOLLOWING FEATURES:

- Lessons are easy to use. Many begin with a sample problem from a real-life experience. After the sample problem is introduced, you are taught step-by-step how to find the answer. Examples show you how to use your skills.
- The ***Guided Practice*** section demonstrates how to solve a problem similar to the sample problem. Answers are given in the first part of the problem to help you find the final answer.
- The ***Exercises*** section gives you the opportunity to practice the skill presented in the lesson.
- The ***Application*** section applies the math skill in a practical or real-life situation. You will learn how to put your knowledge into action by using manipulatives and calculators, and by working problems through with a partner or a group.

Each book ends with ***Cumulative Reviews***. These reviews will help you determine if you have learned the skills in the previous lessons. The ***Selected Answers*** section at the end of each book lists answers to the odd-numbered exercises. Use the answers to check your work.

Working carefully through the exercises in this book will help you understand and appreciate math in your daily life. You'll also gain more confidence in your math skills.

POSSIBLE OUTCOMES FOR A SAMPLE SPACE

Vocabulary

experiment: an activity, such as tossing a number cube or tossing a coin or spinning a spinner

outcome: a result of an experiment

random experiment: an experiment in which each possible outcome is equally likely

sample space: the set of all possible outcomes of an experiment

During the summer, Jesse works for a company that sets up neighborhood carnivals. One of the carnival games uses a spinner which players spin for prizes. What are all the possible **outcomes** when Jesse spins the arrow?

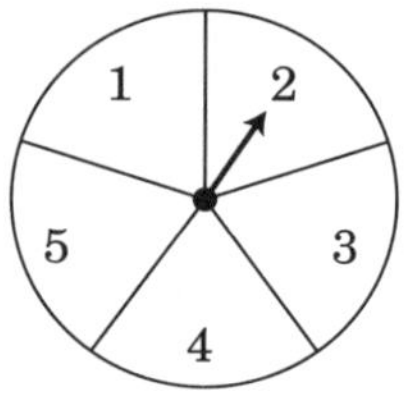

The spinner's arrow can stop on any of the numbers:

1 or 2 or 3 or 4 or 5

So, there are five possible outcomes.

You can write the **sample space** of the game by putting brackets around the set of all possible outcomes as shown below:

{1, 2, 3, 4, 5}

This spinner game is an **experiment**. It is called a **random experiment** because each possible outcome is *equally likely* to happen. This is because the spinner is not weighted and the sections of the spinner are equal in size and shape.

The spinner below has the same sample space as the one above.

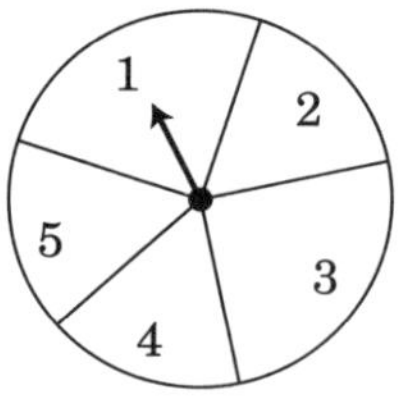

{1, 2, 3, 4, 5}

However, because the sections are different sizes, it is *more likely* that the spinner's arrow would stop on 1 or 3. Therefore, this spinner game would *not* be a random experiment.

Guided Practice

1. Sometimes Jesse uses the spinner shown below.

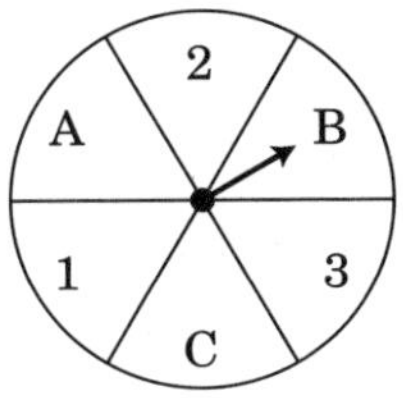

a. List all the possible outcomes for the spinner.

1 or A or 2 or B or 3 or C

b. Write the sample space for the spinner. { _______________ }

c. Are all outcomes equally likely to occur? ___________

d. Would this be a random experiment? ___________

2. In one carnival game, players toss a number cube to win prizes. The sides of the number cube are numbered 1 to 6.

a. List all the possible outcomes. ___________

b. Write the sample space. ___________

c. Would this be a random experiment? Why or why not?

____

3. In another carnival game, a player picks a card out of a hat. The 20 cards in the hat are numbered from 1 to 20.

a. The sample space is ___________.

b. How many possible outcomes are there? ___________

c. Imagine that the player knows that the number 3, 4, and 5 cards have bent corners. Would this be a random experiment? Why or why not? ____

____

Exercises

Write the sample space for each experiment.

4. Tossing a coin ___________

5. Picking an *e* from all the vowels of the alphabet ___________

6. Picking an even number from all the whole numbers 1 to 8

____

7. Spinning the spinner below __________

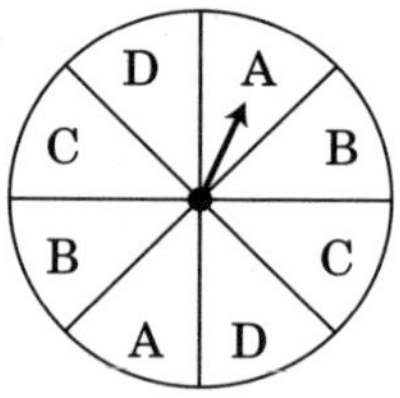

8. Picking the colors of a traffic signal __________

9. Choosing your color when playing checkers __________

10. Tossing a ring onto one of the shapes below __________

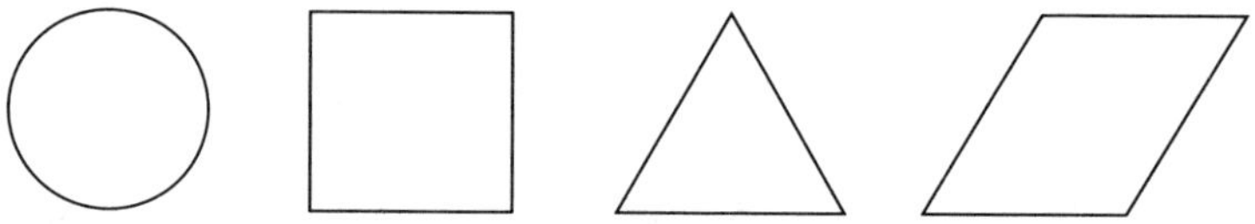

11. Choosing one of four directions on a compass __________

12. Choosing any letter from the word PROBABILITY __________

13. Would Exercise 12 describe a random experiment? Is there an equal likelihood of picking any letter? Explain.

____

____

Application

14. A cleaning crew manager randomly assigns new staff members to cleaning jobs. The jobs are: *vacuum halls, vacuum offices, dust offices, empty wastebaskets,* and *wash windows.*

a. What is the sample space?

____

b. How many possible jobs (outcomes) are there for each new staff member? __________

15. Suppose you have a blank spinner with five equal sections like the one shown below.

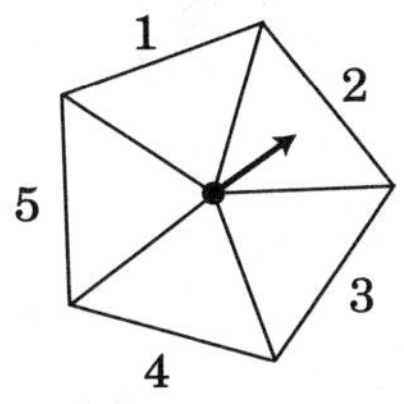

You want to color each section a different color. You choose from the following colors: red, white, blue, green, yellow, and purple.

a. How many different combinations of the colors can you use to color the spinner? ____________

b. Give the sample space for each combination.

____

____

____

____

____

____

c. If the spinner had six equal sections, how would that change the number of sample spaces possible? Explain your answer.

____

____

COOPERATIVE LEARNING

16. Work in a group. Think of a random experiment and describe it below. Have each member of the group find the sample space and compare answers for each experiment.

____

____

THE CONCEPT OF PROBABILITY

Vocabulary

probability: the chance that an event will occur, written as a ratio

favorable outcomes: the number of successful outcomes

Manuel and Luisa are playing a board game. Each player tosses a cube, numbered 1 to 6 to determine the number of spaces to move on the board. To win, Luisa needs to toss a number greater than 4 on her next turn. She wants to know what her chances of winning are, that is, what is the **probability** that she will win?

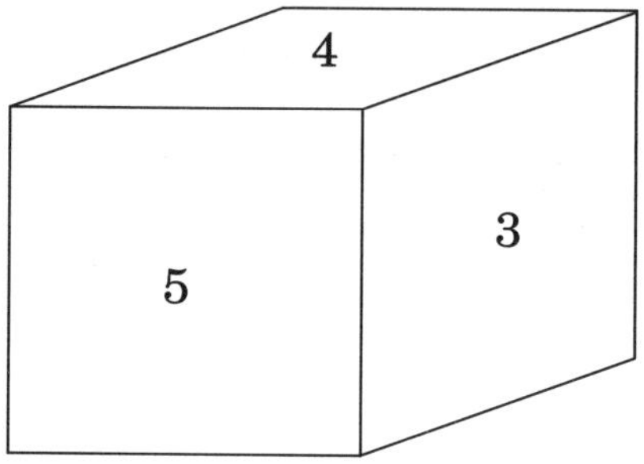

There are 6 possible outcomes when a number cube is tossed. The sample space is {1, 2, 3, 4, 5, 6}.

Two of the possible outcomes, the numbers 5 and 6, are greater than 4.

Luisa finds the *probability* of tossing a number greater than 4 by writing a ratio.

So, there are 2 **favorable outcomes**. The favorable outcomes are the ones that will help Luisa win the game.

$$\frac{\text{Number of favorable outcomes}}{\text{Total number of possible outcomes}} = \frac{2}{6}$$

The probability of tossing a number greater than 4 is 2 out of 6 possible outcomes, or $\frac{2}{6}$.

$$\frac{2}{6} = \frac{2 \div 2}{6 \div 2} = \frac{1}{3}$$

So, Luisa knows that the probability that she will toss a number greater than 4 is $\frac{2}{6}$, or $\frac{1}{3}$.

Reminder

Simplify fractions by dividing the numerator and denominator by the same number.

Guided Practice

Find each probability.

1. On a quiz show, each contestant spins a spinner that is divided into 26 sections. Each section contains one of the 26 letters of the alphabet. Find the probability that the spinner will land on B.

   a. How many spaces are on the spinner? 26

   b. How many spaces have the letter B? ______

   c. What is the probability of landing on the letter B? ______

2. Find the probability that the alphabet spinner will land on a vowel.

   a. How many vowels are on the spinner? ______

   b. What is the probability of landing on a vowel? ______

3. Find the probability that the alphabet spinner will land on a consonant.

   a. How many consonants are on the spinner? ______

   b. What is the probability of landing on a consonant? ______

4. Does a contestant have a greater probability of landing on a vowel or landing on a consonant? Why?

   ______

Exercises

Write the sample space for each experiment and the probability of each outcome.

5. Experiment: Tossing a coin

   a. Sample space ______

   b. Probability of getting heads ______

   c. Probability of getting tails ______

6. Experiment: Spinning this wheel

   a. Sample space ______

   b. Probability of getting the letter M

   ______

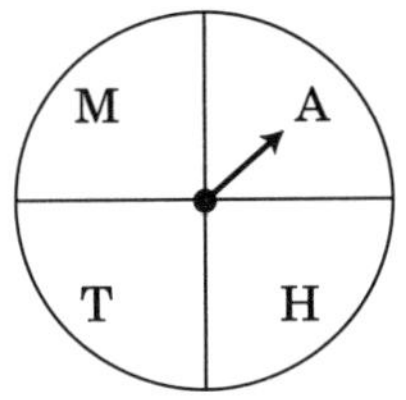

c. Probability of getting the letter T

______

d. Probability of getting a consonant ______

7. Experiment: Tossing a cube numbered 1 to 6

a. Sample space ______

b. Probability of getting a 4 ______

c. Probability of getting an even number ______

d. Probability of getting an odd number ______

e. Which probability is greater, getting an odd number or getting an even number? Explain.

______

8. Experiment: Picking a number between 1 and 10 out of a hat

a. Sample space ______

b. Probability of getting a 1 ______

c. Probability of getting an even number ______

d. Probability of getting an odd number ______

e. Probability of getting a number less than 10 ______

f. Probability of getting a number greater than 5 ______

g. Which probability is greater: getting a number less than 10 or getting a number greater than 5? Explain.

______

Application

COOPERATIVE LEARNING

9. With your group, create a probability experiment. Cut out squares of red, blue, green, and yellow paper or use squares of white paper with the names of the colors written on them. Decide on the total number of squares of paper for each color.

a. Write the sample space. Record how many squares of each color you have.

______

______

b. Find the number of possible outcomes. Then find the probability of choosing each color.

____

____

____

c. Place the squares in a bag. Take 25 turns drawing a square out of the bag. Record each result below. Return the square to the bag each time.

d. Write the final results for each of the outcomes (colors) as a fraction. Write the numbers of times you picked a color as the numerator. Write the total number of times you drew all of the squares as the denominator. Simplify if necessary.

Probability of getting a red is __________

Probability of getting a blue is __________

Probability of getting a green is __________

Probability of getting a yellow is __________

e. The probabilities you wrote before drawing the squares out of the bag are called *theoretical probabilities*. The results of your experiment are called *experimental probabilities*. Why do you think they are different?

____

____

f. Do you think your results would be different if you did the experiment again? Explain.

____

____

3 MORE COMPLEX SAMPLE SPACES

Tanisha has a summer job helping a designer at a department store. She is working on a display of summer wear. There are 2 colors of shorts: blue and red, and 3 colors of shirts: gold, purple, and green. How many different possible combinations of shirts and shorts can Tanisha use in the display?

She makes an organized list of the possible combinations to find out.

Blue shorts, gold shirt

Blue shorts, purple shirt

Blue shorts, green shirt

Red shorts, gold shirt

Red shorts, purple shirt

Red shorts, green shirt

Tanisha can now see that there are 6 possible outcomes in the sample space.

Notice that she organized the list by writing all the blue shorts with shirts combinations first and then repeating for the red shorts combinations.

Guided Practice

1. Zachary and Teresa flip coins. Zachary flips a quarter and Teresa flips a dime.

a. What are the possible outcomes when Zachary flips the quarter?

heads, *tails*

b. What are the possible outcomes when Teresa flips the dime?

__________, __________

c. What are all the possible combinations? Complete the chart below.

quarter heads,	dime heads
quarter ,	dime tails
quarter	
quarter	

2. Ernesto is playing a board game with the two spinners shown below.

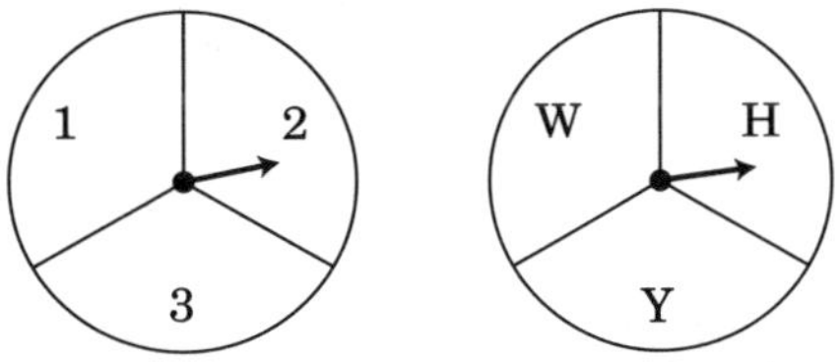

a. List all the possible outcomes for the first spinner.

__________, __________, __________

b. List all the possible outcomes for the second spinner.

__________, __________, __________

c. Use the chart below to show all the possible outcomes when spinning both spinners.

1 W		

d. How many possible outcomes are there altogether? __________

Exercises

Use each chart to list all possible outcomes for each of the experiments.

3. Experiment: Tossing a coin and spinning Wheel 1

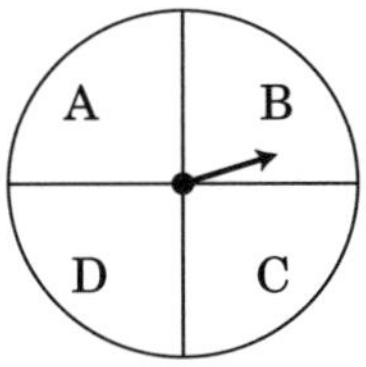

Wheel 1

4. Experiment: Spinning Wheel 2 and Wheel 3

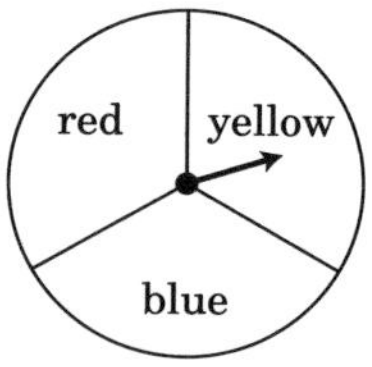

Wheel 2

5. Experiment: Spinning Wheel 1 and Wheel 3

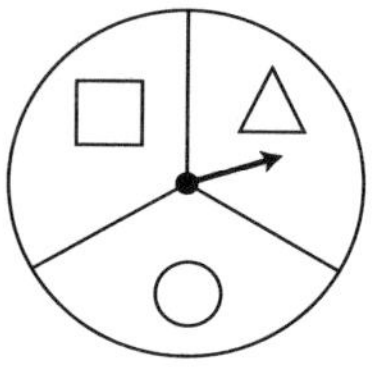

Wheel 3

6. Experiment: Spinning Wheel 1 and Wheel 2

7. Experiment: Spinning Wheel 2, two times

a. How many possible outcomes are there? ____________

b. How many of the outcomes repeat the same color? ____________

c. What is the probability of getting the same color on both spins if you spin Wheel 2 twice? ____________

8. Experiment: Rolling a cube numbered 1 to 6 and spinning Wheel 1.

a. How many possible outcomes are there? ____________

b. How many of the outcomes have the number 6 as one part of the outcome? ____________

c. What is the probability of having a 6 as part of the outcome?

____

9. Experiment: Rolling two cubes, each numbered 1 to 6.

a. How many outcomes are possible? ____________

b. How many of the outcomes have a sum of 7? ____________

c. What is the probability of having an outcome with the sum of 7?

____

Application

10. How can you tell how many possible outcomes there will be in the sample space when two cubes are tossed or two spinners are used?

____

____

4 EVENTS

Vocabulary

event: any possible outcome of an experiment

Example 1

Yoshiko works after school in the receiving department of a hardware store. It is her job to record all inventory received each day. In a shipment of tools, the boxes in the carton are the same size, but they are not labeled. Yoshiko knows that the carton contains 4 boxes of crescent wrenches, 2 boxes of socket wrenches, 1 box of needle-nosed pliers, and 5 boxes of Phillips screwdrivers. If Yoshiko chooses a box at random, what is the probability it will contain crescent wrenches?

The sample space is: {crescent wrenches, socket wrenches, needle-nosed pliers, Phillips screwdrivers}.

The **event**, choosing a box of crescent wrenches, is one of the possible outcomes in the sample space. There are four different events that can occur if Yoshiko chooses a box at random. However, because there is a different number of boxes for each type of tool, the probability of each event occurring is different.

You can use this formula to find the probability of an event.

$$P(\text{event}) = \frac{\text{number of favorable outcomes}}{\text{total number of possible outcomes}}$$

$$P(\text{crescent wrench}) = \frac{\text{4 boxes of crescent wrenches}}{\text{12 total boxes of tools}} = \frac{1}{3}$$

Reminder

Simplify fractions by dividing the numerator and denominator by the same number.

The probability of this event, choosing a box of crescent wrenches, is $\frac{1}{3}$.

Example 2

During her break, Yoshiko plays a game with Jorge. They take turns tossing a coin and rolling a cube numbered 1 to 6. Jorge wants to know the probability of tossing a head and rolling an even number.

Before beginning the game, they find the sample space of all possible outcomes. H–1 means tossing a head on the coin and rolling a 1 on the number cube. T–6 means tossing a tail on the coin and rolling a 6 on the number cube.

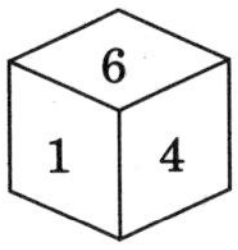

The sample space is

{H-1, H-2, H-3, H-4, H-5, H-6, T-1, T-2, T-3, T-4, T-5, T-6}

Jorge sees that 3 of the 12 outcomes in the sample space, H-2, H-4, H-6, would produce the desired event.

$$P(\text{heads and even number}) = \frac{3}{12} = \frac{1}{4}$$

The probability of this event, tossing a head and rolling an even number, is $\frac{1}{4}$.

Guided Practice

1. Look back at Yoshiko's inventory of the shipment of tools. Find the probability of the other three events.

a. What is the total number of boxes? ____12____

b. How many boxes of socket wrenches are there? __________

c. The probability of choosing a box of socket wrenches is __________.

d. How many boxes of pliers are there? __________

e. The probability of choosing a box of pliers is __________.

f. How many boxes of screwdrivers are there? __________

g. The probability of choosing a box of screwdrivers is __________.

2. In the game that Yoshiko and Jorge play, Yoshiko wants to find the probability of the event of tossing a tail and rolling a number less than 3.

a. List the outcomes that would produce the event that Yoshiko wants.

____

b. How many outcomes are possible? __________

c. What is the probability of this event? ____________

Exercises

Yoshiko is sorting washers of different-size diameters by choosing them from a box at random, one at a time. The box contains 10 $\frac{1}{4}$-in. washers, 15 $\frac{3}{8}$-in. washers, and 12 $\frac{1}{2}$-in. washers.

Find the probability of each event.

3. Choosing a $\frac{1}{4}$-in. washer ____________
4. Choosing a $\frac{3}{8}$-in. washer ____________
5. Choosing a $\frac{1}{2}$-in. washer ____________
6. Choosing a washer with a diameter less than $\frac{1}{2}$ in. ____________
7. Choosing a washer with a diameter greater than $\frac{1}{4}$ in. ____________

Find the probability of each event in the game that Yoshiko and Jorge play.

8. Tossing a head and rolling a 5 ____________
9. Tossing a tail and rolling a 7 ____________
10. Tossing a head and rolling an odd number ____________
11. Tossing a tail and rolling a number greater than 4 ____________

Application

COOPERATIVE LEARNING

12. With your group, create a probability experiment that involves one action (such as tossing a number cube or picking a square of paper from among several squares with different letters on them). Describe your experiment below.

____

____

a. Write the sample space.

____

b. Find the probability of each event.

____

13. Create a second probability experiment in which the events are composed of a combination of *two outcomes*, such as tossing a head and picking a square of paper with an *A* on it. Describe the experiment.

____

____

a. Write the sample space.

____

b. Find the probability of each event.

____

____

5 COMPLEMENT OF AN EVENT

Vocabulary

complement of an event: all possible outcomes other than the given event

certain event: an event which will always occur

impossible event: an event which cannot occur

Reminder

P stands for probability.

Four employees are working together on a team project: Anita, Bashar, Carl, and Daimon. One person will be the recorder. The group decides to choose the recorder by drawing names out of a hat. Anita wonders what her chances are of being chosen as the recorder.

She knows that the number of ways she could be the recorder is 1.

The total number of possible outcomes is 4.

The probability that Anita will be the recorder is $\frac{1}{4}$.

$$P(\text{Anita is the recorder}) = \frac{1}{4}$$

Example 1

What is the probability that Anita will *not* be the recorder?

The number of ways that Anita will not be chosen is 3.

The number of possible outcomes is 4.

$$P(\text{Anita is not the recorder}) = \frac{3}{4}$$

The probability that Anita will *not* be the recorder is $\frac{3}{4}$.

The probability that Anita will *not* be the recorder is the **complement of the event** that Anita will be the recorder.

The sum of the probabilities of all possible outcomes of a given experiment is 1. In the example, Anita's chance of being the recorder was $\frac{1}{4}$ and her chance of *not* being the recorder was $\frac{3}{4}$.

$$\frac{1}{4} + \frac{3}{4} = \frac{4}{4} = 1$$

Another way to find the complement of an event is to subtract the probability of the event from 1.

Example 2

What is the probability that Anita, Bashar, Carl, or Daimon will be the recorder?

$$P(\text{Anita, Bashar, Carl, or Daimon is the recorder}) = \frac{4}{4} \text{ or } 1$$

The probability of a **certain event** is 1.

Example 3

What is the probability that neither Anita, Bashar, Carl nor Daimon will be the recorder?

$$P(\text{not Anita, not Bashar, not Carl, not Daimon}) = \frac{0}{4} = 0$$

The probability of an **impossible event** is 0.

Guided Practice

1. Soon Yi has a bag of marbles: 4 blue, 4 yellow, and 12 red. If she reaches in and picks a marble, what is the probability that she does *not* pick a blue marble?

a. What is the total number of marbles? 20

b. $P(\text{yellow}) = \frac{4}{20} = \frac{1}{5}$

c. $P(\text{red}) =$ ____________

d. $P(\text{not blue}) =$ ____________

e. $P(\text{blue}) =$ ____________

f. $1 - P(\text{blue}) =$ ____________

2. Is the probability that Soon Yi will pick a blue, yellow, or red marble certain or impossible?

a. Number of ways to pick a blue, yellow, or red marble 20

b. Number of ways to pick a marble ____________

c. $P(\text{blue, yellow, red}) =$ ____________

d. P(blue, yellow, red) is certain or impossible? (*choose one*) ____________

3. Is the probability that Soon Yi will pick a purple marble certain or impossible?

a. Number of ways to pick a purple marble ____________

b. Number of ways to pick a marble ____________

c. P(purple) = ____________

d. P(purple) is certain or impossible? (*choose one*) ____________

Exercises

Find the probability of each event.

4. Experiment: Tossing a number cube, numbered 1 to 6

a. Tossing a 2 ____________

b. Not tossing a 2 ____________

c. Tossing a number less than 3 ____________

d. Tossing a number greater than 3 ____________

e. Tossing a 7 ____________

f. Tossing a number less than 7 ____________

5. Experiment: Spinning the wheel

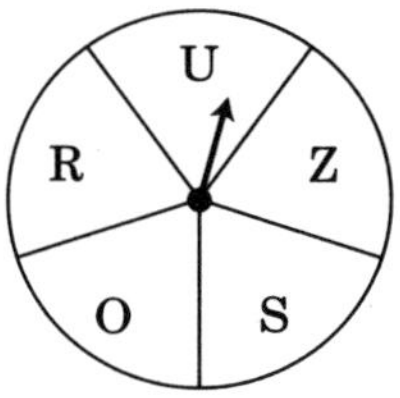

a. Getting a vowel ____________

b. Not getting a vowel ____________

c. Getting a Z ____________

d. Not getting a Z ____________

6. Experiment: Picking a pair of socks from a drawer containing 3 green pairs, 4 blue pairs, 3 red pairs, and 10 white pairs

a. Picking a green pair ____________

b. Picking a red pair ____________

c. Picking a white pair ____________

d. Picking a blue pair ____________

e. Picking a green, red, blue, or white pair ____________

f. Picking a yellow pair ____________

Application

7. Leona works for a company that hires a lot of temporary workers. The company pays overtime to the regular workers when there is no temporary help available. Last year, the company paid overtime during 18 weeks and hired temporary workers for 23 weeks. At the end of the year an auditor randomly selects a week's records. (Assume 52 weeks in a year.)

a. What is the probability that the auditor will check a week in which overtime was paid? ____________

b. What is the probability that the auditor will check a week in which no overtime was paid? ____________

c. What is the probability that the auditor will check a week in which temporary workers were paid? ____________

d. What is the probability that the auditor will check a week in which neither temporary workers nor overtime was paid? ____________

8. Describe a **certain event** in your future.

____

____

9. Describe an **impossible event** in your future.

____

____

6 RANDOM NUMBERS AND PROBABILITY

Vocabulary

random numbers: numbers that are generated in such a way that each digit has a chance of being chosen

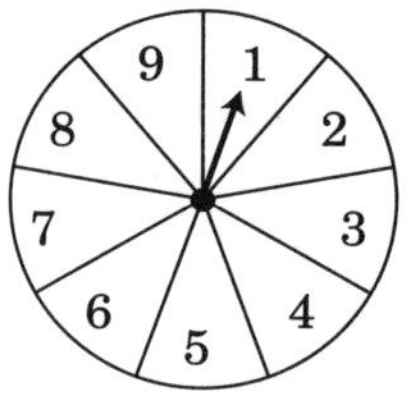

Anna is planning a fund-raiser for the school booster club. She plans to spin a wheel with the digits 1 to 9 to determine the prize winners. She wonders how many prizes she will need if the spinner lands on 5.

Anna uses the spinner above to experiment. She spins the spinner 15 times. These are the numbers that she gets on 15 tries:

2, 3, 9, 4, 8, 7, 1, 4, 7, 7, 1, 2, 5, 7, 3

She notices that there is only *one* 5 in the first *fifteen* spins. Using her list, she sees that the experimental probability of spinning a 5 is

$$P(5) = \frac{1}{15}$$

Next, Anna spins the spinner 50 times and makes this table:

2	3	9	4	8	7	1	4	7	7
1	2	5	7	3	7	8	4	5	9
8	9	4	2	1	3	5	3	2	9
8	1	5	7	1	4	6	3	1	5
2	8	9	5	7	8	6	1	3	2

To find the experimental probability of spinning a 5 using the table of **random numbers**, count the number of 5's in the table.

There are 50 entries and there are six 5's in the table.

$$P(5) = \frac{6}{50}, \text{ or } \frac{3}{25}$$

Guided Practice

1. Use Anna's table of random numbers to find the probability of spinning a 6.

a. Count the 6's in the table. 2

b. Count the entries in the table. ____________

c. $P(6) =$ ____________

2. Use the last five columns of the table to find the experimental probability of spinning a 9.

a. Count the 9's in the 25 entries. ____________

b. $P(9) =$ ____________

Exercises

Use Anna's table of random numbers. Find the experimental probability of each event.

3. Spinning a 4 using the first two rows ____________

4. Spinning a 2 or a 3 using the first two rows ____________

5. Spinning a 7 using the first three rows ____________

6. Spinning a 1, 2, 3, or 4 using the entire table ____________

7. Spinning an even digit using the last row ____________

8. Spinning an even digit using the entire table ____________

9. Spinning an odd digit using the last row ____________

10. Spinning an odd digit using the entire table ____________

Application

With your group, experiment with your calculators to generate random two-digit numbers.

COOPERATIVE LEARNING

11. a. Make a chart of the first 25 random two-digit numbers that you generate.

____

____

b. Using those numbers, find $P(\text{odd numbers}) =$ ____________

c. Compare your results with other groups. How many groups have probabilities of $\frac{12}{25}$ or $\frac{13}{25}$? ____________

RELATIVE FREQUENCY

Vocabulary

frequency: the number of times a particular outcome occurs in an experiment

frequency table: a table listing each outcome and the number of times it occurs

relative frequency: the results of an experiment compared to the possible outcomes; also called the experimental probability

Hinan thinks that 6 is her lucky number. She is playing a board game using a number cube, and she thinks she will toss a 6 more often than any other number. To see if this is true, she performs an experiment.

Hinan tosses a number cube, numbered 1 to 6, 20 times. She makes a **frequency table**, listing the number of outcomes for each of the numbers on the cube.

Experiment 1

Number on the cube	1	2	3	4	5	6
Frequency	3	5	2	4	1	5

Notice that the sum of the frequencies, or the total number of trials, is 20.

$$3 + 5 + 2 + 4 + 1 + 5 = 20 \text{ trials}$$

Hinan's results show that the **frequency** of rolling a 6 is 5.

Reminder

Probability is the ratio of the number of favorable outcomes divided by the number of possible outcomes.

You can find the **relative frequency**, or experimental probability, of tossing a 6 by writing a ratio.

$$\text{Relative frequency} = \frac{\text{Frequency of item}}{\text{Total number of trials}}$$

$$\frac{5 \text{ frequencies}}{20 \text{ trials}} \rightarrow \frac{5}{20} = \frac{1}{4}$$

So, the relative frequency of tossing a 6 is $\frac{1}{4}$.

Guided Practice

Experiment 2

1. Hinan performs the same experiment again. She records the results in the frequency table below.

Number on the cube	1	2	3	4	5	6
Frequency	21	12	15	18	20	14

a. What is the total number of trials? ___100___

b. What is the frequency of tossing a 6? 14

c. Find the relative frequency by writing a ratio.

______

d. Compare the results of Hinan's two experiments.

P(6 for 20 trials) = ______

P(6 for 100 trials) = ______

e. Which probability was greater? ______

Exercises

Ramon conducted an experiment with a spinner. Use his frequency table for Exercises 2 to 6.

Letter on spinner	S	M	I	L	E
Frequency	22	18	14	24	22

2. Find the total number of trials. ______

3. Find the relative frequency of spinning an S. ______

4. Find the relative frequency of spinning an M. ______

5. Find the relative frequency of spinning an L. ______

6. Find the relative frequency of landing on a vowel. ______

Application

7. Place 20 coins in a paper cup. Shake them up and spill them on the table.

a. Record your data in this table.

Number of heads	
Number of tails	

b. What is the relative frequency of heads? ______

c. What is the relative frequency of tails? ______

d. If one coin is thrown on the desk, what is the probability that it will show tails? P (tails) = ______

8 SAMPLING A POPULATION

Vocabulary

population: a large group of people or objects from which a sample is taken

sample: a subgroup that represents the larger population

sampling: collecting data from a sample of the population

A group of environmentalists study the redfish **population** in a pond using a method called "Capture/Recapture." First, they *capture* a **sample** of 300 redfish, tag them, and put them back into the pond. Several weeks later, they *recapture* a sample of 100 redfish. They see that 15 redfish are tagged.

By **sampling** this population, the environmentalists can estimate the number of redfish in the pond, or the size of the population.

First, find the ratio of tagged fish in the original sample of total fish population (*P*).

$$\frac{300}{P}$$

Next, find the ratio of tagged fish in the second sample to the total number of fish in the second sample.

$$\frac{15}{100}$$

Reminder

To solve a proportion, cross multiply and solve for the missing value.

Then, write a proportion and solve.

$$\frac{300}{P} = \frac{15}{100}$$

$$15 \times P = 300 \times 100$$

$$P = 30{,}000 \div 15$$

$$= 2{,}000$$

The environmentalists estimate that there are approximately 2,000 redfish in the pond.

Sampling relies on probability and relative frequency to predict results for larger populations that would be too difficult to count.

Guided Practice

1. To estimate the size of the alligator population in Louisiana swamps, rangers capture 35 alligators, tag them, and then release them. Two months later, rangers recapture 250 alligators, 6 of which are tagged.

a. Write a ratio for the number of alligators tagged to the total population. $\frac{35}{P}$

b. Find the ratio of the tagged alligators to the total number recaptured. ____________

c. Write a proportion. ____________

d. Solve the proportion to estimate the alligator population. $P =$ ____________

Exercises

Estimate the total population using the Capture/Recapture method.

2. Wolves in the forest: Capture and tag 40. Recapture 30, with 8 tagged.

a. Proportion: ____________

b. Estimated population of wolves: ____________

Application

3. Jasmine has a large jar of pennies. She wants to know how many pennies she has, but she doesn't want to count them all. First, she takes 30 pennies from the jar and marks each with a red dot. Next, she places the pennies back in the jar and mixes them. Then, she takes out 30 more pennies, 6 of which have a red dot.

a. Estimate the total number of pennies in Jasmine's jar. ____________

b. Fill a jar with pennies. Repeat Jasmine's experiment. How many pennies are in your jar? ____________

9 PROBABILITY AND PERCENT

Vocabulary

likely: an event probably will occur

not likely: an event probably will not occur

Reminder

Percent means *per hundred.* You can write a percent as a fraction with a denominator of 100.

The Channel 8 weather forecaster reports a 25% chance of rain. Is it **likely** or **not likely** to rain?

Forecasters use the concepts of probability to make their predictions. A 25% chance of rain means rain is 25% likely to occur and 75% not likely.

To change a percent to a probability ratio, write the percent as a fraction.

$$25\% = \frac{25}{100} = \frac{1}{4}$$

$$75\% = \frac{75}{100} = \frac{3}{4}$$

The forecasters know that when weather conditions were similar in the past, it rained 1 out of 4 days and 3 out of 4 days it did not rain. Since $\frac{1}{4} < \frac{3}{4}$, the probability of having rain is less likely than the probability of not having rain.

Guided Practice

1. David hears on TV that there is an 80% chance of a blizzard starting after 2 A.M. What is the probability that there will be a blizzard?

a. Change the percent to a fraction by writing it with a denominator of 100. $80\% = \frac{80}{100}$

b. Simplify the fraction. ____________

c. The probability of a blizzard is ____________.

d. The probability of no blizzard is ____________.

e. Is a blizzard likely or not likely? ____________

Exercises

Write the probability as a ratio. Simplify if possible. Tell if the event is likely or not likely to occur.

2. A 60% chance of snow

____

3. A 45% chance of winning

____

4. A 20% chance of losing home ____________

5. A 50% chance of walking ____________

6. A 75% chance of making a free throw ____________

7. A 5% chance of a penalty ____________

8. A 30% chance of getting a hit ____________

9. A 12% chance of getting a cold ____________

Application

10. Three candidates are running for mayor in the town of Greenville. One week before the election, a survey of voters shows that 23% support Candidate A, 34% support Candidate B, and 41% support Candidate C. If the actual vote follows the survey results, what is the probability that each candidate will win?

a. Candidate A: ____________

b. Candidate B: ____________

c. Candidate C: ____________

d. Should they all stay in the race for mayor? Explain your thinking.

____

____

____

11. A hurricane is traveling up the Eastern seaboard. There is a 60% chance that the storm still will be classified as a hurricane when it reaches Connecticut.

a. Should residents consider leaving the Connecticut shore? __________

b. The storm has a 10% chance of reaching Maine. Do you think authorities will urge residents to leave the Maine shore? __________

c. What other factors might you think about when deciding whether to leave the area because of a possible hurricane?

____

____

____

10 PROBABILITY FOR MUTUALLY EXCLUSIVE EVENTS

Vocabulary

mutually exclusive events: events that have no common outcomes

The Chamber of Commerce is having a raffle to raise money to build a playground for disabled children. It offers winners a pair of tickets to one of the 6 plays, 4 musicals, 3 dance concerts, 2 instrumental concerts, and 5 sports events in town. Winners cannot win more than one prize. If you are a winner, what is the probability that you will win a ticket to a play or a musical?

Winning a ticket to a play and winning a ticket to a musical are **mutually exclusive events** because it is impossible to win both in this raffle.

You can find the probability of mutually exclusive events by finding the sum of their separate probabilities.

First, find the probability of winning tickets to a play. There are 6 plays out of a total of 20 events.

$$P(\text{play}) = \frac{6}{20}$$

Next, find the probability of winning a ticket to a musical. There are 4 musicals out of a total of 20 events.

$$P(\text{musical}) = \frac{4}{20}$$

Reminder

Sometimes it is better not to simplify fractions until after finding their sum.

Then, find the sum of their separate probabilities.

$$P(\text{play or musical}) = \frac{6}{20} + \frac{4}{20} = \frac{10}{20} = \frac{1}{2}$$

The probability of winning a pair of tickets to a play or musical is $\frac{1}{2}$.

Guided Practice

1. Find the probability of winning a pair of tickets to a sports event or a dance concert.

   a. P(sports event) = $\frac{5}{20}$

   b. P(dance concert) = $\frac{3}{20}$

c. P(sports event or dance concert)

$=$ ____________ $+$ ____________ $=$ ____________

d. The probability of winning tickets to a sports event or a dance concert is ____________.

2. Find the probability of *not* getting tickets to a play by adding the probabilities of every outcome except a play.

a. P(musical) $=$ ____________

b. P(dance concert) $=$ ____________

c. P(instrumental concert) $=$ ____________

d. P(sports event) $=$ ____________

e. P(not play) $=$ ____________ $+$ ____________ $+$ ____________ $+$ ____________ $=$ ____________

Reminder

The complement of an event is all possible outcomes other than the given event.

3. Find the probability of not getting tickets to a play.

a. P(play) $=$ ____________

b. P(not play) $= 1 - P$(play) $=$ ____________

Exercises

4. Alice has two cubes, each numbered 1 to 6. If she tosses them both at the same time, what is the probability that she will roll a sum of 4 or 9?

a. Complete the table of possible outcomes.

(1–1)	(1–2)	(1–3)	(1–4)		
(2–1)					
(3–1)					
(4–1)					

b. P(sum of 4) $=$ ____________

c. P(sum of 9) $=$ ____________

d. P(sum of 4 or 9) $=$ ____________

5. Use the table above to determine each probability.

   a. P(sum of 7 or 11) = ____________

   b. P(sum of 1 or 12) = ____________

   c. P(sum of 3 or 5) = ____________

   d. P(sum of 7 or 2) = ____________

   e. P(sum of 6 or 8) = ____________

   f. P(even sum or odd sum) = ____________

6. Experiment: Flip two coins at the same time.

   a. P(2 heads) = ____________

   b. P(2 tails) = ____________

   c. P(1 tail and 1 head in any order) = ____________

   d. P(no heads) = ____________

7. Experiment: Spin a wheel with the numbers 1 to 10, in equal sections.

   a. P(number less than 3 or greater than 8) = ____________

   b. P(a multiple of 5 or multiple of 7) = ____________

   c. P(an odd number or even number) = ____________

   d. P(number greater than 10) = ____________

8. Several local restaurants worked together to boost sales. They held a drawing in Oak Forest for a free family meal. There are 2 Italian, 3 Indian, 4 Chinese, 2 Thai, and 2 Mexican restaurants in town. If you have a winning ticket, what is the probability of winning a meal at an Indian or Mexican restaurant? ____________

Application

COOPERATIVE LEARNING **Work with a partner.**

9. Place some squares of colored paper in a bag. Use four different colors and at least 25 squares.

   a. Each partner lists all the possible outcomes and the probability of each individual event. Compare your lists before you proceed.

   b. Take turns drawing a square out of the bag and replacing it before the next person draws. Do this at least 25 times. Fill in the frequency chart to show the results.

Color 1:	Color 2:	Color 3:	Color 4:

c. Make a list of at least three events that are mutually exclusive. Use your frequency chart and add the results together to find the relative frequency of these mutually exclusive events.

d. Compare the relative frequency of the mutually exclusive events that you found in **c**, above, to the theoretical probability of each event that you described in **a**, above.

e. Explain the differences you found between the theoretical probability and the relative frequency of each of the mutually exclusive events.

10. Try the experiment again. How do your results differ from your first experiment? What do you think will happen if you repeat the experiment 10 more times?

TREE DIAGRAMS

Vocabulary

tree diagram: a diagram that shows all possible outcomes for an experiment

At Petra's Restaurant, there is a special deal on Tuesday nights. For a low price, a customer can have a three-course dinner with the following choices: 1) black bean soup or onion soup, 2) steak, flounder, or quiche, and 3) cake or pie. How many different meal combinations are possible?

To determine all the possible meal combinations, Petra made a **tree diagram**.

Tuesday Night Menu Choices

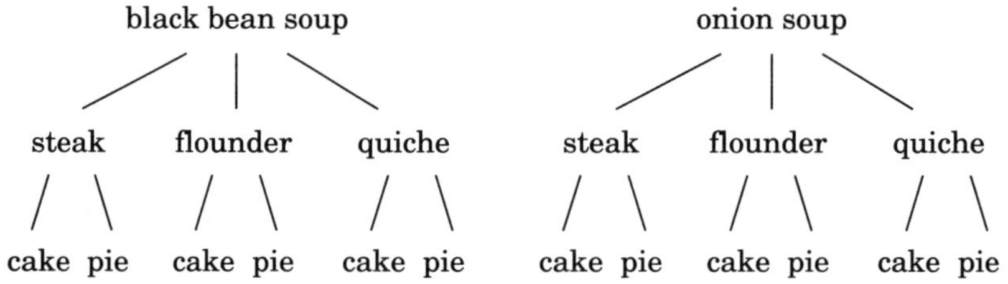

To use the diagram, start from the top of each "tree." Follow each branch to find every possible meal. Make a list.

1. *black bean soup, steak, and cake*
2. *black bean soup, steak, and pie*
3. *black bean soup, flounder, and cake*
4. *black bean soup, flounder, and pie*
5. *black bean soup, quiche, and cake*
6. *black bean soup, quiche, and pie*
7. *onion soup, steak, and cake*
8. *onion soup, steak, and pie*
9. *onion soup, flounder, and cake*
10. *onion soup, flounder, and pie*

11. onion soup, quiche, and cake

12. onion soup, quiche, and pie

There are 12 different meal combinations available at Petra's Restaurant on Tuesday nights.

To find the probability of any combination, divide a specific menu by the total possibilities.

P(black bean soup, flounder, pie) = $\frac{1}{12}$

Guided Practice

1. The new chef has suggested that Petra's Restaurant offer a choice of three desserts: cake, pie, or ice cream.

a. Complete this tree diagram to show all possible meal combinations.

NEW Tuesday Night Menu Choices

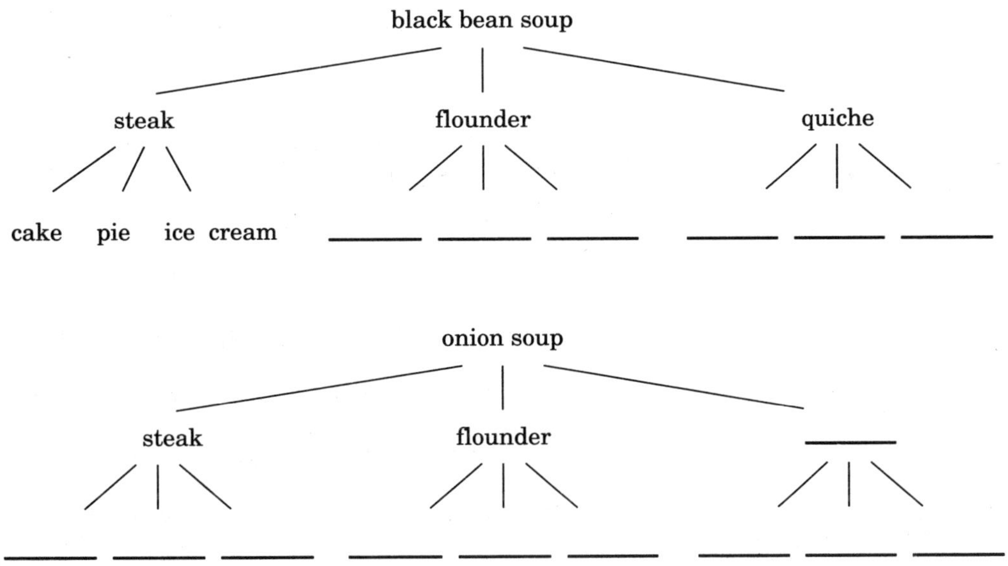

b. Count each branch in the bottom rows of both tree diagrams to find out how many meal combinations are now possible.

____

c. To find the probability of any given menu, divide the number of ways to choose that specific menu by the total number of menus.

P(onion soup, steak, ice cream) = __________

Exercises

Draw a tree diagram for each experiment. Count how many possible combinations or outcomes there are in each case.

2. Flipping two coins ______________ outcomes

3. Flipping three coins ______________ outcomes

4. Making a three-digit code using the numbers 4, 5, and 6 (each digit may be used more than once) ______________ outcomes

5. Rolling a cube numbered 1 to 6, and then tossing a coin ______________ outcomes

6. Rolling two number cubes, each numbered 1 to 6 ______________ outcomes

Use the tree diagrams from Exercises 1 to 6 to find the probabilities for the following outcomes.

7. P(1 head and 1 tail) when flipping two coins ________

8. P(3 heads) when flipping three coins ________

9. P(2 heads and 1 tail) when flipping three coins ________

10. P(number 4 and 1 head) when rolling a number cube and tossing a coin ________

11. P(sum of 4) when rolling two number cubes ________

12. P(getting a three-digit code containing exactly two 4's) ________

13. P(getting a three-digit code containing exactly two 5's) ________

14. P(getting a three-digit code containing exactly one 4) ________

15. P(getting a three-digit code without any 4's) ________

Application

COOPERATIVE LEARNING

Work with a partner.

16. At Thurgood Marshall Middle School, the physical education classes are sponsoring a series of sports-related activities.

- Students can choose from the following activities: 100-yard dash, javelin throw, bicycle race, standing broad jump, and pole vault.
- The students can take a written test about physical fitness in English, Spanish, or Vietnamese.
- Students can choose to have the student teacher or the master teacher evaluate their overall performances.

a. Make a tree diagram to determine the possible combinations of events, testing, and evaluation options.

b. If students are given a choice of an oral or written test, how many combinations of events, testing, and evaluation options are possible?

____

12 INDEPENDENT EVENTS

Vocabulary

independent events: two events with outcomes that do not depend on each other

Andre rolls a six-sided cube, numbered 1 to 6 , and tosses a coin. He wonders about the probability of getting a 4 on the cube and heads on the coin.

Rolling a cube has no effect on tossing a coin. The two are **independent events.**

To find the probability, he makes a table of all the possible outcomes.

H1	H2	H3	H4	H5	H6
T1	T2	T3	T4	T5	T6

Andre looks at the table and finds that the probability of getting a 4 and heads is $\frac{1}{12}$.

$$P(4 \text{ and } H) = \frac{1}{12}$$

If two events A and B are independent, the probability that both events occur is the probability of A multiplied by the probability of B.

$$P(\text{A and B}) = P(\text{A}) \times P(\text{B})$$

The probability of getting heads on a coin toss is $\frac{1}{2}$. The probability of rolling a 4 is $\frac{1}{6}$.

Andre uses the formula above because he knows he has independent events.

$$P(4 \text{ and } H) = \frac{1}{6} \times \frac{1}{2} = \frac{1}{12}$$

The results are the same.

Guided Practice

1. What is the probability that Andre will get tails and an odd number?

   a. Use the table to list the favorable outcomes.

   ___T1___, ________, ________

   b. Total number of possible outcomes = ________

   c. *P*(T and odd) = ________

   d. *P*(T) = ________ and *P*(odd) = ________

   e. Using the formula, *P*(T and odd) =

   ________ × ________ = ________

2. A booth at the school carnival has two game wheels. One has equal areas of red, blue, yellow, and green. The other has equal areas numbered 1, 2, 3, and 4. The most valuable prize requires spins of blue on one wheel and 1 on the other. What is the probability of spinning blue and 1?

   a. To find the probability, complete the table for all possible outcomes.

R1			
R2			
R3			
R4			

   b. *P*(blue and 1) = ________

   c. *P*(blue) = ________ and *P*(1) = ________

   d. Using the formula, *P*(blue and 1) =

   ________ × ________ = ________.

Exercises

3. Make a table of all the possible outcomes for tossing a six-sided cube, numbered 1 to 6, and spinning a four-color wheel containing white, black, purple, and orange sectors of equal area. Find the probability for each event listed below.

   a. *P*(even number and purple) ________

   b. *P*(black and number less than 5) ________

   c. *P*(white and a positive number) ________

   d. *P*(3 and orange) ________

   e. *P*(a color and 5) ________

   f. *P*(purple and a number divisible by 3) ________

Use the formula *P*(A and B) = *P*(A) × *P*(B) to find the probability for each event.

4. Tossing coins.

   a. Tossing 3 coins and getting 3 heads ________

   b. Tossing 2 coins and getting 2 tails ________

5. Rolling double 6's on two number cubes ________

Application

6. Nabil has two number cubes, with four sides numbered 1, 2, 3, and 4.

   a. What is the probability that he will roll two even numbers when throwing the two dice? ________

   b. What is the probability that he will roll two numbers, each less than 4? ________

7. Emilita works at a day care center after school. She is planning a card game to teach the children to read their numbers and the picture cards.

a. Get a deck of playing cards, or make one with cut paper and red and black markers.

b. Lay out the cards as Emilita would for the children.

Clubs	x	x	x	x	x	x	x	x	x	x	x	x	x
Diamonds	x	x	x	x	x	x	x	x	x	x	x	x	x
Hearts	x	x	x	x	x	x	x	x	x	x	x	x	x
Spades	x	x	x	x	x	x	x	x	x	x	x	x	x
	2	3	4	5	6	7	8	9	10	J	Q	K	A

c. A child draws a card at random. What is the probability that the card is a club? ____________

d. The child puts the card back. What is the probability that the card is a king? ____________

e. The child puts the card back. What is the probability that the card is the king of clubs? ____________

3 DEPENDENT EVENTS

Vocabulary

dependent events: a set of events in which the outcome of the first event affects the outcome of the next event

Jasmine and her friends are planning a card trick. She shuffles the cards at the right and puts them face down on the table.

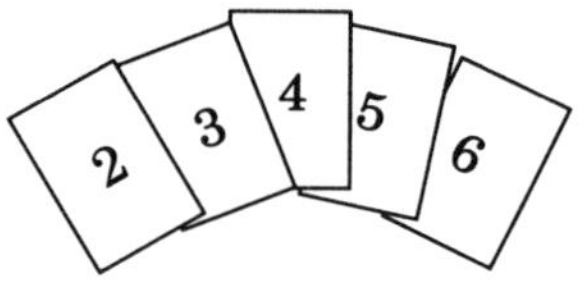

Jasmine asks, "If I pick a card, what is the probability of picking a 3?" Her friends answer quickly, "One in five," or $\frac{1}{5}$.

$$P(3) = \frac{1}{5}$$

"Okay," Jasmine says. "What if I don't replace the first card? What is the probability of picking a 2 as the second card?"

Jasmine points out that because the first card has been chosen, they will have to find the probability of picking the 2 from the four remaining cards. She shows this outcome as "one in four," or $\frac{1}{4}$.

$$P(2) = \frac{1}{4}$$

Seeing that her friends do not understand, Jasmine explains that the outcome of picking the first card affects the outcome of picking the second card. The probability of drawing the 2 as the second card is dependent upon drawing the 3 as the first card. For that reason, these are called **dependent events.**

Jasmine shows her friends how to figure the probability of the two dependent events.

$$P(3, \text{ then } 2) = \frac{1}{5} \times \frac{1}{4} = \frac{1}{20}$$

So, the probability of picking a 3 as the first card and a 2 as the second card is $\frac{1}{20}$.

Guided Practice

1. Martin has ten pieces of paper with one of the digits 0 to 9 on each. He picks three digits, one at a time without looking, to make a three-digit code. What is the probability that Martin will pick the code 852?

a. The probability of an 8 on the first pick is $\frac{1}{10}$.

b. How many digits remain in the sample space? ______

c. The probability of a 5 on the second pick is ______.

d. How many digits remain in the sample space? ______

e. The probability of a 2 on the third pick is ______.

f. Multiply the three probabilities.

___ × ___ × ___ = ___

g. $P(852) =$ ______

2. If the following cards are face down and Mary picks two cards, find the probability of picking two aces (ace, ace).

ace, queen, 10, 8, ace, ace, queen, 10

a. What is the probability of picking an ace as the first card?

$P(\text{ace}) = \frac{3}{8}$

b. Once the first card is set aside, what is the probability of picking an ace as the second card?

$P(\text{ace}) = \frac{2}{7}$

c. $P(\text{ace, then ace}) =$ ______

Exercises

Walden has a bag with 5 yellow counters, 3 blue counters, and 2 green counters. He picks one counter at a time and does not replace it. Find the probability of each of the following.

3. Picking 2 yellow counters.

$P(\text{Y, then Y}) =$ ______

4. Picking a yellow counter, then a blue counter.

$P(\text{Y, then B}) =$ ______

5. Picking a green counter, then a blue counter.

$P(\text{G, then B}) =$ ______

6. Picking a yellow counter, then a green counter, then a blue counter.

P(Y, then G, then B) = ____________

Suppose Walden adds 6 red counters and 4 white counters to the bag. Find the probability of each of the following. Use your calculator, if necessary.

7. Picking 3 counters: red, then blue, then red.

P(R, then B, then R) = ____________

8. Picking 2 green counters, then a white counter.

P(G, then G, then W) = ____________

9. Picking 3 counters: white, then red, then yellow.

P(W, then R, then Y) = ____________

Application

10. Tom and his friends Nick, Steve, Juan, and Aiden are waiting to get into a movie. Just as they get to the ticket window, the manager puts up a sign saying there are only two seats left, one in the first row and one next to the exit sign. All five boys put their school identification cards into Tom's hat and shake it up. Juan pulls out two names, the first for the first-row seat and the second for the exit sign seat.

   a. What is the probability that Tom gets the first-row seat and Aiden gets the exit-sign seat? ____________

   b. Make a list of all the possible outcomes.

   Start your list this way:

   Tom, Nick

   Tom, Steve

   Tom, Juan

   Tom, Aiden

c. How does your answer for P(Tom, then Aiden) compare with your list of outcomes?

____

d. Suppose Steve decides to go home. So the other four put their identification cards into the hat to decide who will see the movie. What is the probability that Juan and Aiden get to see the movie?

____

11. Make a set of cards with the following letters.

A, B, C, D, E, F, G, H, I, J

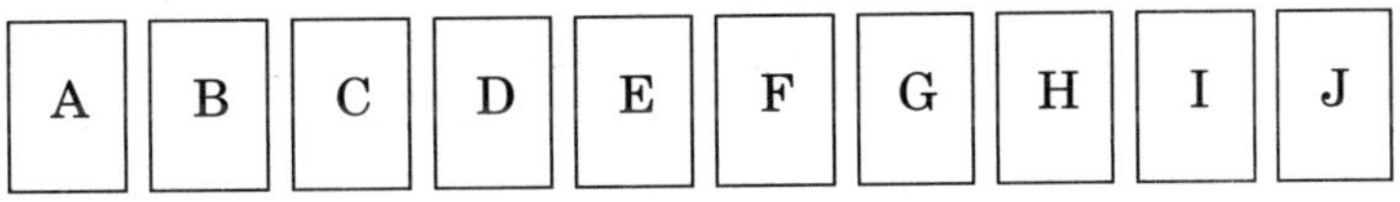

Find the probability of each of the following picks if the first card picked is set aside.

a. What is P(A, then B)? P = __________

b. What is P(A, then vowel)? P = __________

c. What is P(vowel, then vowel)? P = __________

d. What is P(vowel, vowel, consonant, vowel). Use your calculator.

P = __________

GEOMETRIC PROBABILITY

Hannah has a quilt with light squares and dark squares. The quilt contains 30 squares, each 9 inches on a side. If a drop of coffee lands on Hannah's quilt, what is the geometric probability that it will land on one of the dark squares?

To find the probability, find the total area of the dark squares, the total area of the quilt, and write a ratio.

$$P(\text{dark square}) = \frac{\text{Area of the dark squares}}{\text{Area of the quilt}}$$

Reminder

To find the area of a rectangle, multiply length × width. Area is measured in square inches, or in.^2

Find the area of the dark squares.

$$\text{Area} = 14 \text{ squares} \times (9 \text{ in.} \times 9 \text{ in.}) = 1{,}134 \text{ in.}^2$$

Find the total area of the quilt.

$$\text{Area} = 30 \text{ squares} \times (9 \text{ in.} \times 9 \text{ in.}) = 2{,}430 \text{ in.}^2$$

Write a ratio of dark squares to total area, and simplify.

$$P(\text{dark square}) = \frac{1{,}134}{2{,}430} = \frac{7}{15}$$

So, the geometric probability that a coffee drop will land on a dark square is $\frac{7}{15}$.

Guided Practice

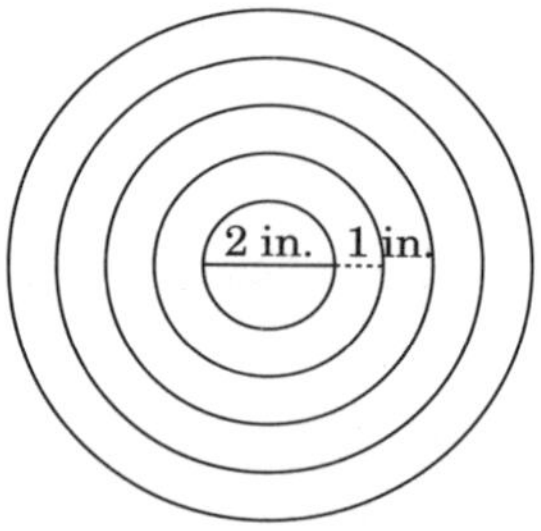

1. There is a dartboard in the student lounge. It has five concentric circles, alternating black and gold. The bull's-eye (center circle) is 2 inches in diameter and each surrounding ring is 1 inch wide. What is the probability of getting a bull's-eye by hitting the center circle?

Reminder

The radius is $\frac{1}{2}$ of the diameter of a circle.

a. Find the area of the bull's-eye. (Use $\pi = 3.14$.)

$A = \pi r^2 = \pi \times 1^2 =$ 3.14 sq. in.

b. Find the radius of the dartboard by adding the radius of the center to the width of the four surrounding rings.

Radius of the dartboard = 5 in.

c. Find the area of the dartboard.

$A = \pi \times r^2 = 3.14r^2 =$ ____________

d. P(bull's-eye) = Area of bull's-eye / Area of dartboard =

____________ = ____________ / ____________

Exercises

A checkerboard has 8 rows of 8 squares, with 4 red squares and 4 black squares in each row. Each square measures 1 inch on a side. Find the geometric probability of each result if a checker is dropped onto the board. Draw a diagram of the checkerboard to help you.

2. Landing on a red square on a checkerboard.

a. Write a ratio. ____________

b. P(red square) = ____________

3. Landing on one of the squares along the outside edge of the checkerboard. Count the corner squares one time only.

a. Write a ratio. ____________

b. P(outside square) = ____________

Application

4. Suppose each square measures 2 inches on a side. Will that affect the geometric probability of landing on a red square? Explain your reasoning.

____

____

5. A tennis court is 78 feet long. It is 27 feet wide for singles tennis and 36 feet wide for doubles tennis. If a fan throws a ball down to the court, what is the probability that it will land in the area of the court used only for doubles? ____________

5 PERMUTATIONS

Vocabulary

permutation: an arrangement of a group of items in which order is important

Miguelina Perez is an electrical inspector for the city of Dalton. She must inspect three sites: the library, the post office, and the town hall. To plan her day, she decides to find out in how many orders she could visit these sites.

A specific arrangement of the items in a group is called a **permutation.** Miguelina makes a table to find the number of permutations possible for the three sites she must inspect. Each row represents a different order.

First site	Second site	Third site
library	post office	town hall
library	town hall	post office
post office	town hall	library
post office	library	town hall
town hall	post office	library
town hall	library	post office

She finds there are six different orders, or permutations, for the three sites she must inspect.

You can use the counting principle to find all the permutations for three items taken three at a time.

3 Choices for the 1st site	**2** Choices for the 2nd site	**1** Choice for the 3rd site

$$3 \times 2 \times 1 = 6$$

Guided Practice

1. Suppose Miguelina also has to inspect a museum that day. Find the possible number of permutations of four inspection sites, taking four at a time.

   a. Make a table or a list of all the possible permutations.

   b. In how many different orders can Miguelina visit four inspection sites? ______

   c. How many permutations of four things, taken four at a time, are possible? Use the counting principle.

   ______ × ______ × ______ × ______ = ______

Exercises

Find the number of permutations for each situation.

2. Choosing three letters from L, M, Q.
3. Choosing two colors from red and blue.
4. Arranging a wrench, hammer, screwdriver, and saw on a tool rack.
5. Choosing five numbers from 1, 2, 3, 4, and 5.
6. Arranging the letters in the word TEAM, using all the letters each time.
7. Arranging seven books on a shelf.
8. Which of the following are permutations: credit-card numbers, members of a committee, license-plate numbers?

Application

COOPERATIVE LEARNING

9. Find the number of permutations of five classes that are possible, taken four at a time.
   - **a.** With your group, list five classes (English, math, science, etc.).
   - **b.** Find how many different groups of four classes each are possible using the five options.
   - **c.** Find the number of permutations for each group of four classes.
   - **d.** How many permutations of five classes, taken four at a time, are possible? List the possibilities.

6 COMBINATIONS

Vocabulary

combination: a selection of items in which order is not important

Cliff Lonetree makes Native American jewelry. He uses the following stones in his jewelry: coral, turquoise, amber, and jet. In any ring, he uses two different types of stones. Cliff wants to know how many different ways he can combine two of the four types of stones.

Cliff made a list of all possible **combinations** of two stones.

coral-turquoise	turquoise-coral
coral-amber	amber-coral
coral-jet	jet-coral
turquoise-amber	amber-turquoise
turquoise-jet	jet-turquoise
amber-jet	jet-amber

Because order does not matter in a combination, Cliff finds that the pairs in the second column are the same as those in the first column. So, he can only make six different combinations of two from the four stones available.

Guided Practice

1. Cliff wants to use a combination of two different stones in a necklace. He is using coral, turquoise, and amber stones.

   a. Complete the combinations that use coral.

   coral, *turquoise*　　　coral, *amber*

   b. Complete the combination for turquoise that Cliff has not used yet.

   turquoise, ______

   c. Can any other combinations be made that have not been used already? ______

   d. How many different combinations of two stones can Cliff make from the three types of stones?

   ______

Exercises

Find the number of combinations.

2. Choosing two metals from gold, silver, and pewter.

____

3. Choosing three items from ring, bracelet, necklace, and earrings.

____

4. Choosing four shapes from square, circle, triangle, trapezoid, and diamond.

____

5. Choosing three colors from red, blue, green, black, and white.

____

6. Think about the following: telephone numbers, inventory list, social-security numbers, student class list. Which of these are combinations?

____

Application

COOPERATIVE LEARNING

7. Work with a partner to make a list of five items.

a. ____

____

____

b. Choose the number of items you want in your combination.

____

____

c. List all possible combinations that you could make using your five items. How many combinations are possible?

____

____

8. Describe the process you used to find all possible combinations for your items.

____

____

____

LESSONS

1-3 CUMULATIVE REVIEW

Write the sample space for each experiment.

1. Experiment: Choosing any letter from the word "VACATION"

____

2. Experiment: Picking a red marble from a bag containing 20 marbles—10 marbles are red, 5 are green, 3 are yellow, and 2 are blue

____

3. Experiment: Ordering a baked potato with possible toppings of butter, sour cream, chives, bacon bits, cheese, and onions

____

Answer each question.

4. When is an experiment called a random experiment?

____

5. How many outcomes of an experiment are listed in a sample space?

____

6. A spinner has ten equal sections labeled 1 to 10. What is the probability that you will spin a 6 or an 8?

____

7. If you draw a card from a deck of 52 bridge cards, what is the probability you would get an ace?

____

8. Make a table to show all possible outcomes of rolling a six-sided number cube and tossing a coin.

9. Bob has a new job and wants to wear a different combination of clothes every day. He has two pairs of slacks, navy and gray, and four shirts: white, gray, blue, and yellow. Make a list of Bob's possible outfits.

____

LESSONS

4-6 CUMULATIVE REVIEW

1. Roland spins the spinner. Find the probability of each event.

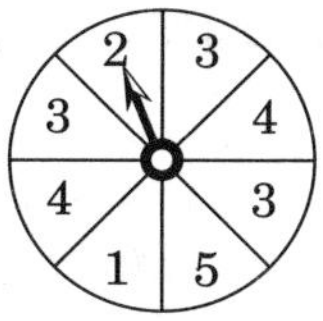

a. Spinning a 2 ____________

b. Spinning an odd number ____________

c. Spinning a number greater than 3 ____________

d. Spinning a 6 ____________

e. Spinning a number from 1 through 5 ____________

f. Not spinning a 4 ____________

2. Marc tosses two coins. Find the probability of each event.

a. P(2 tails) = ____________

b. P(head first, tail second) = ____________

c. P(no tails) = ____________

d. P(one head and one tail) = ____________

3. Irene has five cards labeled A, E, I, O, U in one box and three cards labeled 3, 6, 9 in a second box. She picks one letter card and then one number card. Find the probability of each event.

a. P(A, 6) = ____________

b. P(O, 9) = ____________

c. P(E, 4) = ____________

d. Picking the letter A, E, I, O, or U and the number 3 ____________

e. Picking the letter I and the number 3, 6, or 9 ____________

4. Sanchita has a cube that is numbered 1 to 6. She rolls the cube 12 times. These are the numbers she gets: 2, 3, 5, 6, 1, 3, 2, 2, 6, 5, 1, 2. Use the list to find the experimental probability of each event.

a. P(5) = ____________

b. P(4) = ____________

c. P(1, 2, or 3) = ____________

d. P(number greater than 4) = ____________

e. P(number less than 7) = ____________

LESSONS
7-9 CUMULATIVE REVIEW

Use the frequency table below to answer the following:

Letter on spinner	P	A	R	T
Frequency	31	19	22	28

1. Find the frequency of spinning an A. ____________
2. Find the frequency of spinning a T. ____________
3. Find the frequency of spinning a consonant. ____________
4. Find the relative frequency of spinning an A. ____________
5. Find the relative frequency of spinning a T. ____________
6. Find the relative frequency of spinning a consonant. ____________
7. For which letters is the relative frequency greater than the theoretical probability? ____________

Predict the total population using the Capture/Recapture method.

8. Sparrows in Central Park: Capture and tag 50. Recapture 30, with 6 tagged. ____________
9. Fish in Lake Flounder: Capture and tag 100. Recapture 100, with 25 tagged. ____________
10. Groundhogs in Montreal: Capture and tag 400. Recapture 200, with 40 tagged. ____________

Write each probability as a ratio. Tell if the event is likely or unlikely to occur.

11. A 33% chance of winning a marathon ____________
12. A 68% chance of being selected ____________
13. A 60% chance of scoring a goal ____________
14. A 25% chance of a tornado ____________

LESSONS

10-11 CUMULATIVE REVIEW

1. A deck of playing cards has 52 cards. A student draws a card at random. After each choice, the student returns the card to the deck. Find each probability.

   a. P(red card) = ____________

   b. P(ace) = ____________

   c. P(red card or ace) = ____________

   d. P(6 or 7) = ____________

2. Draw a tree diagram to show all possible outcomes of flipping four coins.

   ____________ outcomes

3. A new restaurant opened and offers two soups, three entrées, two styles of potatoes, and four desserts. The soups are onion and chicken noodle; the entrées are beef, chicken, and fish; the potatoes are mashed and baked; and the desserts are cake, pie, ice cream, and rice pudding.

   ____________ outcomes

4. Experiment: Toss 3 coins. Find each probability.

   a. P(2 heads and 1 tail in any order) = ____________

   b. P(at least 2 tails) = ____________

LESSONS 12-14 CUMULATIVE REVIEW

1. Make a table showing the sample space for spinning a wheel with four equal sections labeled E, A, S, T, two times. Use your table to find the probability of each event below.

   a.

   b. P(S, then T) ____________ d. P(any two vowels) ____________

   c. P(two T's) ____________ e. P(at least one vowel) ____________

Find the probability of each of the following.

2. Rolling two 2's using two number cubes. ____________
3. Rolling one 3, then one 4, using two number cubes. ____________
4. Getting a 3 and a tail by rolling a 6-sided number cube and tossing a dime. ____________
5. Getting a head and a 5 or a 6 by tossing a coin and rolling two number cubes. ____________
6. Rolling an odd number and a number greater than 2, using two number cubes. ____________

Six cards are in a box. They are numbered 4, 5, 6, 7, 8, and 9. Find each probability if the cards are drawn and set aside.

7. Drawing the number 45. ____________
8. Drawing the number 789. ____________
9. Drawing two odd numbers. ____________
10. Drawing an odd number and a 6. ____________
11. The ninth grade must elect a president and a vice president for the class. Adam, Edith, Giorgio, James, and Rosita are running for office. The student with the most votes will become president, and the runner-up will become vice president. What is the probability that Rosita will become president and James will become vice president?

    ____

LESSONS

15-16 CUMULATIVE REVIEW

Find the number of permutations for each situation.

1. Seating 3 people in 3 seats. ____________
2. Arranging 4 magazines in a rack. ____________
3. Arranging the letters A, B, C, D, E. ____________
4. Choosing 2 ties from 2 ties. ____________
5. Arranging 7 bikes in a row. ____________
6. Picking 3 digits from 2, 3, 4, 5. ____________

Find the number of combinations for each situation.

7. Choosing 3 players from 4 players. ____________
8. Picking 2 CDs from 5 CDs. ____________
9. Selecting 4 numbers from 33, 39, 45, 51, 65. ____________
10. Picking 3 cards from 6 cards. ____________
11. Selecting 2 colors from purple, orange, green, black. ____________
12. In how many different ways can 5 friends be paired off? ____________

ANSWER KEY

LESSON 1 (pp. 2–5)

1. a. 1 or A or 2 or B or 3 or C
b. {1, 2, 3, A, B, C} **c.** Yes **d.** Yes

3. a. {1, 2, 3, 4, 5, 6, 7, 8, 9, 10, 11, 12, 13, 14, 15, 16, 17, 18, 19, 20} **b.** 20 **c.** No, because the cards with the bent corners could be identified.

5. {a, e, i, o, u} **7.** {A, B, C, D}

9. {red, black} **11.** {north, south, east, west}

13. No, because it is more likely that B or I would be chosen than any of the other letters.

15. a. 6
b. {red, white, blue, green, yellow}
{red, white, blue, green, purple}
{red, white, blue, yellow, purple}
{red, white, green, yellow, purple}
{red, blue, green, yellow, purple}
{white, blue, green, yellow, purple}
c. Only one sample space would be possible because all colors would be used.

LESSON 2 (pp. 6–9)

1. a. 26 **b.** 1 **c.** $\frac{1}{26}$

3 a. 21 **b.** $\frac{21}{26}$

5 a. {heads, tails} **b.** $\frac{1}{2}$ **c.** $\frac{1}{2}$

7. a. {1, 2, 3, 4, 5, 6} **b.** $\frac{1}{6}$ **c.** $\frac{3}{6}$ or $\frac{1}{2}$
d. $\frac{3}{6}$ or $\frac{1}{2}$ **e.** The probability of getting an odd number is equal to the probability of getting an even number. Both probabilities are $\frac{1}{2}$.

9. a.–f. Answers will vary by group.

LESSON 3 (pp. 10–13)

1. a. heads, tails **b.** heads, tails

c.

quarter heads, dime heads
quarter heads, dime tails
quarter tails, dime tails
quarter tails, dime heads

3.

heads - A	heads - B	heads - C	heads - D
tails - A	tails - B	tails - C	tails - D

5.

A - square	A - triangle	A - circle
B - square	B - triangle	B - circle
C - square	C - triangle	C - circle
D - square	D - triangle	D - circle

7.

red - red	red - yellow	red - blue
yellow - red	yellow- yellow	yellow - blue
blue - red	blue - yellow	blue - blue

a. 9 **b.** 3 **c.** $\frac{3}{9} = \frac{1}{3}$

9.

1 - 1	2 - 1	3 - 1	4 - 1	5 - 1	6 - 1
1 - 2	2 - 2	3 - 2	4 - 2	5 - 2	6 - 2
1 - 3	2 - 3	3 - 3	4 - 3	5 - 3	6 - 3
1 - 4	2 - 4	3 - 4	4 - 4	5 - 4	6 - 4
1 - 5	2 - 5	3 - 5	4 - 5	5 - 5	6 - 5
1 - 6	2 - 6	3 - 6	4 - 6	5 - 6	6 - 6

a. 36 **b.** 6 **c.** $\frac{6}{36} = \frac{1}{6}$

LESSON 4 (pp. 14–17)

1. a. 12 **b.** 2 **c.** $\frac{2}{12}$ or $\frac{1}{6}$ **d.** 1 **e.** $\frac{1}{12}$
f. 5 **g.** $\frac{5}{12}$

3. $\frac{10}{37}$ **5.** $\frac{12}{37}$ **7.** $\frac{27}{37}$ **9.** $\frac{1}{12}$ **11.** $\frac{2}{12}$ or $\frac{1}{6}$

13. a.–b. Check group experiments.

LESSON 5 (pp. 18–21)

1. a. 20 **b.** $\frac{4}{20} = \frac{1}{5}$ **c.** $\frac{12}{20} = \frac{3}{5}$ **d.** $\frac{16}{20} = \frac{4}{5}$
e. $\frac{4}{20} = \frac{1}{5}$ **f.** $1 - \frac{1}{5} = \frac{4}{5}$

3. a. 0 **b.** 20 **c.** $\frac{0}{20} = 0$ **d.** Impossible

5. a. $\frac{2}{5}$ **b.** $\frac{3}{5}$ **c.** $\frac{1}{5}$ **d.** $\frac{4}{5}$

7. a. $\frac{18}{52} = \frac{9}{26}$ **b.** $\frac{34}{52} = \frac{17}{26}$ **c.** $\frac{23}{52}$ **d.** $\frac{11}{52}$

9. Answers may vary. Sample: living on Mars

LESSON 6 (pp. 22–23)

1. a. 2 **b.** 50 **c.** $\frac{2}{50} = \frac{1}{25}$

3. $\frac{3}{20}$ **5.** $\frac{5}{30} = \frac{1}{6}$ **7.** $\frac{5}{10} = \frac{1}{2}$ **9.** $\frac{5}{10} = \frac{1}{2}$

11. a. Any 25 two-digit numbers
b. $\frac{\text{(number of odds)}}{25}$ **c.** Answers will vary.

LESSON 7 (pp. 24–25)

1. a. 100 **b.** 14 **c.** $\frac{14}{100} = \frac{7}{50}$ **d.** $\frac{1}{4}, \frac{7}{50}$
e. P (6 for 20 trials)

3. $\frac{22}{100} = \frac{11}{50}$ **5.** $\frac{24}{100} = \frac{6}{25}$

7. a. Check students' tables.
b. $\frac{\text{number of heads}}{20}$ **c.** $\frac{\text{number of tails}}{20}$ **d.** $\frac{1}{2}$

LESSON 8 (pp. 26–27)

1. a. $\frac{35}{P}$ **b.** $\frac{6}{250} = \frac{3}{125}$ **c.** $\frac{35}{P} = \frac{6}{250}$
d. 1458.3 or about 1,458 alligators

3. a. 150 pennies **b.** Check students' experiments.

LESSON 9 (pp. 28–29)

1. a. $\frac{80}{100}$ **b.** $\frac{4}{5}$ **c.** $\frac{4}{5}$ **d.** $\frac{1}{5}$ **e.** likely

3. $\frac{45}{100} = \frac{9}{20}$; not likely

5. $\frac{50}{100} = \frac{1}{2}$; equal chance

7. $\frac{5}{100} = \frac{1}{20}$; not likely

9. $\frac{12}{100} = \frac{3}{25}$; not likely

11. a. yes **b.** no **c.** Answers may vary. Sample: Road conditions, intensity of wind and rain, the speed the storm is traveling, whether it's day or night, if the area is urban or rural, are other factors you might think about before leaving the area.

LESSON 10 (pp. 30–33)

1. a. $\frac{5}{20}$ **b.** $\frac{3}{20}$ **c.** $\frac{5}{20} + \frac{3}{20} = \frac{8}{20} = \frac{2}{5}$ **d.** $\frac{2}{5}$

3. a. $\frac{6}{20}$ **b.** $1 - \frac{6}{20} = \frac{14}{20} = \frac{7}{10}$

5. a. $\frac{6}{36} + \frac{2}{36} = \frac{8}{36} = \frac{2}{9}$ **b.** $\frac{1}{36}$
c. $\frac{2}{36} + \frac{4}{36} = \frac{6}{36} = \frac{1}{6}$ **d.** $\frac{6}{36} + \frac{1}{36} = \frac{7}{36}$
e. $\frac{5}{36} + \frac{5}{36} = \frac{10}{36} = \frac{5}{18}$ **f.** $\frac{18}{36} + \frac{18}{36} = \frac{36}{36} = 1$

7. a. $\frac{4}{10} = \frac{2}{5}$ **b.** $\frac{3}{10}$ **c.** $\frac{10}{10} = 1$ **d.** 0

9. a.–e. Answers will vary.

LESSON 11 (pp. 34–37)

1. a.

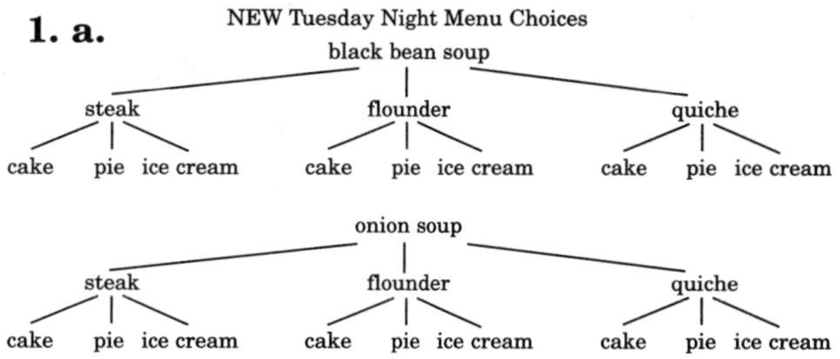

b. 18 **c.** *P*(onion soup, steak, ice cream) $= \frac{1}{18}$

3. Coin toss

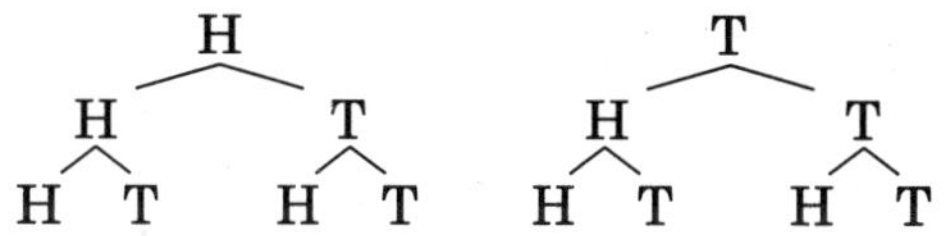

8 outcomes: HHH, HHT, HTH, HTT, THH, THT, TTH, TTT

5.

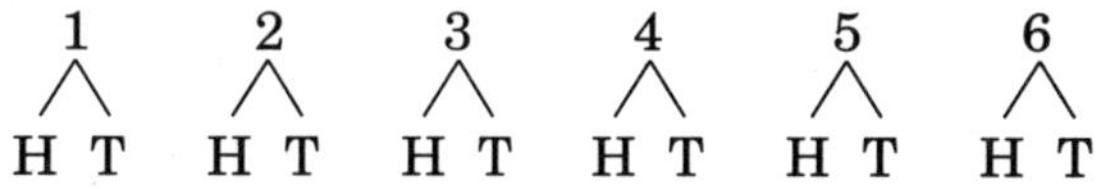

12 outcomes: 1H, 1T, 2H, 2T, 3H, 3T, 4H, 4T, 5H, 5T, 6H, 6T

7. $\frac{2}{4} = \frac{1}{2}$ **9.** $\frac{3}{8}$ **11.** $\frac{3}{36} = \frac{1}{12}$ **13.** $\frac{6}{27} = \frac{2}{9}$

15. $\frac{8}{27}$

LESSON 12 (pp. 38–41)

1. a. T1, T3, T5 **b.** 12 **c.** $\frac{3}{12} = \frac{1}{4}$ **d.** $\frac{1}{2}; \frac{3}{6}$
e. $\frac{1}{2} \times \frac{3}{6} = \frac{3}{12} = \frac{1}{4}$

3. a. $\frac{3}{24} = \frac{1}{8}$ **b.** $\frac{4}{24} = \frac{1}{6}$ **c.** $\frac{6}{24} = \frac{1}{4}$ **d.** $\frac{1}{24}$
e. $\frac{4}{24} = \frac{1}{6}$ **f.** $\frac{2}{24} = \frac{1}{12}$

5. $\frac{1}{36}$

7. c. $\frac{13}{52} = \frac{1}{4}$ **d.** $\frac{4}{52} = \frac{1}{13}$ **e.** $\frac{1}{52}$

LESSON 13 (pp. 42–45)

1. a. $\frac{1}{10}$ **b.** 9 **c.** $\frac{1}{9}$ **d.** 8 **e.** $\frac{1}{8}$
f. $\frac{1}{10} \times \frac{1}{9} \times \frac{1}{8} = \frac{1}{720}$ **g.** $\frac{1}{720}$

3. $\frac{5}{10} \times \frac{4}{9} = \frac{20}{90} = \frac{2}{9}$ **5.** $\frac{2}{10} \times \frac{3}{9} = \frac{6}{90} = \frac{1}{15}$

7. $\frac{6}{20} \times \frac{3}{19} \times \frac{5}{18} = \frac{1}{76}$ **9.** $\frac{4}{20} \times \frac{6}{19} \times \frac{5}{18} = \frac{1}{57}$

11. a. $\frac{1}{10} \times \frac{1}{9} = \frac{1}{90}$ **b.** $\frac{1}{10} \times \frac{2}{9} = \frac{2}{90} = \frac{1}{45}$
c. $\frac{3}{10} \times \frac{2}{9} = \frac{6}{90} = \frac{1}{15}$
d. $\frac{3}{10} \times \frac{2}{9} \times \frac{7}{8} \times \frac{1}{7} = \frac{42}{5{,}040} = \frac{1}{120}$

LESSON 14 (pp. 46–47)

1. a. $3.14 \times (1 \text{ in.})^2 = 3.14$ sq in.

b. $\frac{1}{2} \times 2$ in. + 1 in. + 1 in. + 1 in. + 1 in. = 5 in. **c.** $3.14 \times (5 \text{ in.})^2 = 3.14 \times 25$ sq in.^{2} = 78.5 sq in.^{2} **d.** $\frac{3.14}{78.5} = \frac{1}{25}$

3. a. $\frac{\text{Area outside squares}}{\text{Area checkerboard}}$ **b.** $\frac{28}{64} = \frac{7}{16}$

5. $36 - 27 = 9$; $9 \times 78 = 702$

$36 \times 78 = 2{,}808$

$\frac{702}{2{,}808} = \frac{1}{4}$

LESSON 15 (pp. 48–49)

1. a. Check students' work.

b. 24 different orders

c. $4 \times 3 \times 2 \times 1 = 24$

3. 2 **5.** 120 **7.** 5,040

9. a. a list of any five classes **b.** 5 **c.** 24

d. 120; Answers will vary.

LESSON 16 (pp. 50–51)

1. a. turquoise, amber **b.** amber **c.** no

d. 3

3. 4 **5.** 10

7. a. Check students' lists. **b.** Any number from 2 to 5 **c.** Check combinations.

For 5 items — 1 combination

For 4 items — 5 combinations

For 3 items — 10 combinations

For 2 items — 10 combinations

CUMULATIVE REVIEW (L1–L3) p. 52

1. {V,A,C,T,I,O,N}

3. {butter,sour cream,chives,bacon bits,cheese,onions}

5. All possible outcomes **7.** $\frac{4}{52}$, or $\frac{1}{13}$

9. Navy slacks, white shirt

Navy slacks, gray shirt

Navy slacks, blue shirt

Navy slacks, yellow shirt

Gray slacks, white shirt

Gray slacks, gray shirt

Gray slacks, blue shirt

Gray slacks, yellow shirt

CUMULATIVE REVIEW (L4–L6) p. 53

1. a. $\frac{1}{8}$ **b.** $\frac{5}{8}$ **c.** $\frac{3}{8}$ **d.** 0 **e.** 1 **f.** $\frac{6}{8}$, or $\frac{3}{4}$

3. a. $\frac{1}{15}$ **b.** $\frac{1}{15}$ **c.** 0 **d.** $\frac{5}{15}$, or $\frac{1}{3}$ **e.** $\frac{3}{15}$, or $\frac{1}{5}$

CUMULATIVE REVIEW (L7–L9) p. 54

1. 19 **3.** 81 **5.** $\frac{28}{100}$, or $\frac{7}{25}$ **7.** P and T **9.** 400

11. $\frac{33}{100}$; unlikely **13.** $\frac{60}{100} = \frac{3}{5}$; likely

CUMULATIVE REVIEW (L10–L11) p. 55

1. a. $\frac{26}{52} = \frac{1}{2}$ **b.** $\frac{4}{52} = \frac{1}{13}$

c. $\frac{26}{52} + \frac{4}{52} = \frac{30}{52} = \frac{15}{26}$ **d.** $\frac{4}{52} + \frac{4}{52} = \frac{8}{52} = \frac{2}{13}$

3.

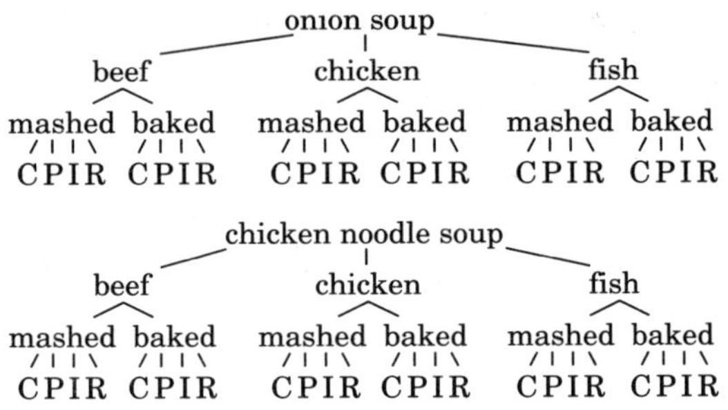

Key

c = cake

p = pie

i = ice cream

r = rice pudding

48 outcomes

CUMULATIVE REVIEW (L12 - L14) p. 56

1. a.

E - E	A - E	S - E	T - E
E - A	A - A	S - A	T - A
E - S	A - S	S - S	T - S
E - T	A - T	S - T	T - T

b. $\frac{1}{16}$ **c.** $\frac{1}{16}$ **d.** $\frac{4}{16}$, or $\frac{1}{4}$ **e.** $\frac{12}{16}$, or $\frac{3}{4}$

3. $\frac{1}{36}$ **5.** $\frac{20}{72}$, or $\frac{5}{18}$ **7.** $\frac{1}{30}$ **9.** $\frac{1}{5}$ **11.** $\frac{1}{20}$

CUMULATIVE REVIEW (L15 - L16) p. 57

1. 6 **3.** 120 **5.** 5,040 **7.** 4 **9.** 5 **11.** 6